Space Psychology

Written by:
Austin Mardon, Catherine Mardon, Monica Thakadu,
Botho Modutlwe, Zeest Kadri, Maya Nagorski,
Kirithika Bharatselvam, Zain Kadri, & Brianna Bedran

Edited by:
Anastasiya Yermolenko

GM PRESS

Typeset and Cover Design by Kim Huynh

ISBN: 978-1-77369-644-7
EBook ISBN: 978-1-77369-645-4

Golden Meteorite Press
103 11919 82 St NW
Edmonton, AB T5B 2W3
www.goldenmeteoritepress.com

Table of Contents

Chapter 1

Origins of Space Psychology including Theories and Discoveries

Monica Thakadu

Introduction

There is a great connection between space, psychology as well as theories and discoveries more than you can ever imagine. This implies that our minds, feelings, thoughts and emotions usually influence how we humans interact with the outside world; be it the earth, moon or space in general and vice versa. In this section we will explore space psychology in detail as well as the key innovators, contributors, theories and beliefs and how their advancement shaped our understanding in modern days.

Origins of Space Psychology

One may ask him/herself questions like what, when, why and who started space psychology. Well space psychology refers to the use of the basics of psychology; i.e. the scientific study of thoughts and mental processes; to advise human spaceflight, [Space Psychology]. It is believed that the study of space or space age began on October 14, 1957 in which the world's first artificial satellite orbiting the Earth was created by the Soviet Union. As more advanced technologies came into life, theories and practical means by Konstantin Tsiolkovsky and Robert H. Goddard led to the discovery of the first liquid fueled rocket [History of space flight].

The works of Edwin Hubble and cosmonaut Yuri Gagarin also contributed to the rise of space psychology. Edwin Powell Hubble was the first human to think of space. He classified galaxy systems, Robert Smith [Expanding universe-1982]. This actually explains

1

why Edwin Hubble is often regarded as "The man who discovered the cosmos." Studies show that indeed Hubble influenced the way we thought about space. Yuri Gagarin on the other hand was "the first man to go to space." He was a Soviet pilot and cosmonaut who got into space using the most powerful rocket at that time.

With all that has been said above, space psychology came to being with the help of other scientists, inventors, contributors, and innovators as it has even expanded our knowledge nationwide. This also includes the evolution of psychologists that attend to astronauts, cosmonauts and other people visiting space as it has also been revealed that people go to space for different reasons; either to expand or improve scientific research or for greatest life achievements ever made in history.

Key Innovators and Contributors in Space Psychology

There are a wild range of milestones of innovations and contributions in these fields of space psychology. Many people were eager to learn about space, the moon and what actually took place out there. These include among others the works of Neil Armstrong, Robert H. Goddard, Eugene A. Cernan, Buzz Aldrin, Charles "Pete" Conrad, David Scott, John Young and Valentina Tereshkova. Not forgetting Harrison Jack Smith, Edgar Mitchell and Edwin P. Hubble. Most of these people were motivated by the works of the National Aeronautics and Space Administration commonly known as NASA. This shows that According to Douglas A. Vakoch[August 7, 2017] NASA is the main contributor of space psychology though the Soviet Union also played a role. Now let's just look into the key innovators and the contributors in full detail.

Key Innovators

Robert H. Goddard: Even though Robert H. Goddard is not that famous in the field of science we cannot talk of space psychology and leave him behind. Robert Hutchings Goddard [October 5,1882 - August 10, 1945] introduced the world's first liquid fueled rocket on March 16 ,1926. He came up with better tactics that enabled rockets to control their flight more effectively. That is why after his death due to Tuberculosis; he was regarded as the "Father of American Rocketry". If you remember quite very well you would realize that

2

it was the period in which the "space age era" began. This therefore further tries to explain why he was also called the "Man who ushered in the space age," [History of spaceflight].

Neil Armstrong: The most famous person in space psychology is Neil Armstrong. He made history under his name by becoming "The first man to walk to the moon." This took place on July 20, 1969 [The first person to the moon] in which he left his footprint on the surface of the moon. It is said that after he walked on the moon he quoted the following words, "That's one small step for man, one giant leap for mankind." This was during the Apollo 11 mission inspired by National Aeronautics and Space Administration [NASA].Armstrong died at the age of 82. Due to Neil Armstrong's achievement ever made in history many other scientists, astronauts and researchers were inspired to try their best of luck and reach the moon too. We will also look into them.

The 12 Men to Walk on the Moon

A) **Buzz Aldrin:** Aldrin who is now 89 became the second man to walk to the moon. He was an Apollo 11 astronaut who stood next to the lunar module "Eagle" on July 20, 1969. He was with Neil Armstrong at that time. Aldrin also quoted some words whilst on the moon, "Beautiful view. Magnificent desolation." Aldrin suffered symptoms of altitude sickness and migrated from the south pole in December 2016 in which he quickly recovered, [the 12 men to walk to the moon – NBC News].

B) **Charles "Pete" Conrad:** Charles Conrad who died at the age of 69 due to a motorcycle accident was the third man to walk to the moon on November 19, 1969. Unlike the first two people who set foot on the moon, Charles quoted the following words, "Whoopee! Man, that may have been a small one for Neil, but that's a long time for me." He was in the Apollo 12 mission together with Alan Bean who resigned from NASA in June 1981 and died at the age of 86 on May 26, 2018, [the 12 men to walk to the moon-NBC News].

C) **Edgar Mitchell and Alan Shepard:** Mitchell who died at the age of 85 carried out a seismic experiment during the 14th Apollo mission to the moon on February 5, 1971. He was eagerly dedicated to studying the mind, physics, psychics and aliens. Alan Shepard on the other hand died in 1988 and was the first American to use the suborbital flight in 1961 before walking on the moon, [the 12 men to walk to the moon-NBC News].

D) James Irwin and David Scott: James A. Irwin, who resigned from NASA and Air force in July 1972 and died in 1991 was a lunar module pilot in the Apollo 15 mission on August 1, 1971. But there was something unique about David Scott; he was the only one amongst the 3 astronauts that has flown both the earth orbital and lunar Apollo missions. He is 87 by the way [the 12 men to walk to the moon-NBC News]. They both used the lunar rover simulator to explore a lava flow at the base of the Sierra Nevadas in California on April 29, 1971.

E) Charles Duke and John Young: Charles Duke who is now 83 made history by being the youngest man to walk on the moon. He was 36 years old at that time. Duke retired from NASA in 1973. Now let's talk about what John Young did. He was the 9th man to walk on the moon and was typically the only agency astronaut to go to space as part of the Gemini, Apollo and space shuttle programs. The interesting part is that he was the first person to fly into space six times. He died at the age of 87 on January 5, 2008. This was the Apollo 16 crew.

F) Harrison Jack Schmitt and Eugene A. Cernan: Harrison Jack Schmitt who is 84 years old was the first person initially trained as a scientist "Geologist" to walk on the moon. Actually NASA chose him together with other scientist-astronauts-the first group not to be test pilots. Eugene A. Cernan was the "last man to stand on the moon"; actually it was the last 17th Apollo mission to be held by NASA ever in history. Cernan travelled into space three times and to the moon twice as pilot of Gemini 9A in June 1966, as lunar module pilot of Apollo 10 in May 1969 and as commander of Apollo 17 in December 1972, the final Apollo lunar landing, [Gene Cernan-Wikipedia].

Edwin P. Hubble: As I have mentioned earlier Edwin Powell Hubble [1889-1953] was the very first man to think of space. He classified galaxy systems as elaborated by Robert Smith[expanding universe-1982]. This actually explains why Edwin Hubble was regarded as the "man who discovered the cosmos." His findings helped to tell us that the universe is expanding.

Yuri A. Gagarin: Yuri Alekseyevich Gagarin [March 9, 1934-March 27, 1968] was a Soviet pilot and a cosmonaut who got into space using a Vostok rocket, the most powerful rocket at that time. He was selected by the Soviet Union to be "the first human into space" on April 12, 1961 under Vostok 1 space mission and was 27 years at that time. Just like most astronauts, Gagarin also quoted his words, "An astronaut cannot be suspended in space and not have God in his mind and his heart," [first man enters space].

Contributors

Now let's look at the contributors to space psychology which are National Aeronautics and Space Administration [NASA] and the Soviet Union. They kind of have the same vision which is selecting astronauts that are well fit enough to go to space or the moon or orbit around.

NASA: The main contributor of space psychology is the National Aeronautics and Space Administration [NASA]. It was formed to deeply find how psychology genuinely plays a role in space race, orbiter and international space station missions and space journeys yet to come [what does NASA stand for?].

A) Robert Goddard: There was little support from people and because he was a shy person he decided to work on his own privately. After his death to honor his hard work NASA's Goddard Space Flight Centre was created in 1959 and also nominated for the International Aerospace Hall of Fame in 1966 and the International Space Hall of Fame in 1976, Michael Foale [22 June 2016].

B) Edwin P. Hubble: It was in his honor that NASA named "The Hubble Space Telescope" after his name[story about The Hubble Space Telescope/NASA].

C) NASA was the coordinator of the Apollo space missions most importantly Apollo 11 to Apollo 17. It carefully selected astronauts and scientists to go on space journeys looking at their strengths and weaknesses [Psychology of space exploration].

Advancement of Theories and Beliefs in Space Psychology, Including How Past Theories Have Shaped our Understanding Today

Theories

Theories are usually essential in our day to day lives in order for us to understand issues beyond our human understanding. Just like any other concept Space Psychology is also characterised by theories that surround it. This includes among others the Theory of Relativity, Field Theory and Mental Space Theory.

Theory of Relativity: The Theory of Relativity which is commonly known as the Theory of Relativity, motion and gravity is the type of theory in space psychology that tries to explain the interaction or the relationship between space, time, gravity and movement. According to Lucian Gideon Conway III, Meredith A. Repke, Shannon C. Houck[November 1, 2016] this theory literally implies that humans going to space should legally consider slowing down time as movement speeds up. It also implies that time and human movement generally affect the way we logically think. Therefore movement through space generally affects movement through time and vice versa.

Field Theory: The Field Theory tries to explain the relationship of patterns of association between a certain individual and the space; Field Theory [psychology] Wikipedia. It tries to explain to us how a certain astronaut, cosmonaut or any researcher interacts with other people; his /her companions and to the environment as well. This looks further into an individual's personal traits and skills like communication skills, anger management skills and time management skills. This came about by a German Psychologist, Kurl Lewin.

Mental Space Theory: Unlike other two theories the Mental Space Theory tries to explain the relationship between an individual's intellectual being and the space. The way the individual thinks; will they be able to think logically outside the box if anything happens. It is sometimes called Mental Theory and Misunderstanding. This one came by Gilles Fauconnier in 1985.

Beliefs

There has been an out rise of beliefs in space psychology but most importantly for astronauts. From a long time ago in the early 1950s there has been the belief in sexism. Many people believed that women were not eligible enough to work in some jobs such as the military and airforce. They believed that only men were eligible to work those jobs for society. Now let's look deeply into the issue. We are going to look into the issue of Sally Ride, Valentina Tereshkova and Sharon Christa McAuliffe; some of the women who made history in space psychology

Valentina Tereshkova: Valentina Tereshkova was the very first woman to go to space in 1963 and spent about three full days there; Karl Tate [June 14, 2013]; a cosmonaut by profession.

Sharon Christa Mcauliffe: Was an American teacher and astronaut who died on a mission of Space Shuttle Challenger where she was a Payload Specialist on January 28, 1986. Her goal in the challenger was to carry out an experiment and to teach two lessons.

Sally Ride: Sally Ride was the first American woman to go to space in 1983 decades after Valentina Tereshkova made history as the world's first woman to go to space; [History.com].

Now let's look into the Sexism part; the part of Gender Inequality.

In the case of Valentina Tereshkova though women qualified to be astronauts with Engineering Degrees and graduates from Jet Pilot testing they were not allowed in the military. Even the 37 women who passed the NASA barrage selection test none of them were allowed to go to space. John Glenn even supported the fact that women are not allowed to participate in the Aerospace journey.

In the case of Sally Ride, yes she was able to go to space but even though she was congratulated by many there was some criticism. When she applied to NASA to be a well known Astronaut she got an unpleasant response; "Your willingness to serve your country as a volunteer is commendable. However, we have no present plans to employ women in space flight because of the degree of scientific and flight training, physical characteristics which are required." Even the media asked her critical questions like how will she manage the issues relating to menstrual period and putting on bras in space.

The question remains; why would people criticize women?

The fact still remains women are very capable to do any work either in the mines, military or airforce.

How Past Theories Shaped Our Understanding Today

The Theories have taught us that anyone who is willing to go to space should be:

• Mentally fit and a logical thinker
• An individual with good interpersonal skills like communication , anger management and listening skill
• Punctual and manage time well, L.G Conway III [Nov 1, 2016]

Conclusion

The Space Age and the innovations, inventions and contributions by scientists, astronauts, cosmonauts, the National Aerospace and Space Administration as well as the Soviet Union all gave birth to what we call Space Psychology. Field Theory, Theory of Relativity, motion and gravity and Mental Space Theory are the past theories that played an important role in selecting well fit astronauts and cosmonauts into space journeys and also has changed our understanding about space Psychology in modern days. Sexism has been a general belief across the ages in space psychology. There is therefore a need to stand up against it. Anyone who has the passion to be an astronaut, cosmonaut or work in the military air force should be able to follow their dreams without any form of criticism. There should be no gender inequality. Everyone is capable of doing whatever they want as long as they are well fit enough and willing to do so. Therefore no one has the right to prevent someone from achieving even their childhood dreams just because they are women. Any individual who is willing to go to space should have a good interactive relationship , logical thinking and be able to control movement and time against gravity whilst in space to avoid any necessary accidents.

Chapter 2

Mental health and the human mind in space

Botho Modutlwe

In this chapter we will look into a broad picture of what mental health is and the factors that influence it. We will describe how the human mind functions in order to maintain the good health of astronauts and cosmonauts including the parts of the brain which are responsible for maintaining a good state of mind. This will also contain a firm description of salutogenic experience and earth photography, eudaimonic well-being, and the environmental challenges and the psych of being in space. With addition of how astronauts survive and thrive in space.

Mental health is a state of well-being in which an individual realizes his or her own abilities, can cope with the normal stresses of life, can work productively, manage the stresses of life, and is able to make contributions to his or her own community, (Adam Felman 2020). The human mind (brain) consists of several parts that have been studied in relation to mental health. These parts include the amygdala, the prefrontal cortex, the anterior cingulate cortex, and hippocampus. These parts of the brain carry out different tasks that help to manage and control the state of the individual's mental health in various fields of mental processing, (7activestudio 2017). The important section relating to the minds of people in space being the prefrontal cortex which is linked to the amygdala. It is responsible for controlling the response to fearful, stressful events, decision making and problem solving. Astronauts and Cosmonauts in space are expected to be able to handle the challenges that come with setting up from the earth and surviving in space. They should be able to hold onto the pressure and the fears and anxieties they may face in space so as to avoid panic attacks that may lead to errors and affect their survival in space. However all these parts of their brain should be in an effective position to help them think properly and process the events going on in space.

Astronauts are more worried about screwing up than blowing up in space, (The Royal Institution, 2015). According to Dr. Sarita Robinson, there are different identified examples of problems that astronauts may face when they are in space which are much bigger than when one is on earth. Robinson says the fact that astronauts need to be able to maintain a high amount of stress response for a long period of time, can lead to problems like exhaustion, inter-group conflicts and the workload that astronauts and cosmonauts have to complete. These include a lot of experiments with very little or no rest which may be difficult for the individual to say they are anxious or are experiencing some element of depression. It is much easier for them to cover up those issues as they are more focused on getting their work done.

Yes, surviving in space can be really challenging and demands one to be in a good mental, emotional, spiritual and physical state. One has to be in good shape to withstand the environment of space and its features that are so extremely different from earth. The environment can be exciting and you don't want to get too exhilarated in space as you might lose control of yourself and end up losing yourself. Also you should not be too scared of the weird features of space, self-control is highly crucial and for that your mind should be under a very stable position.

Salutogenesis Experience (Earth Photography)

A medical sociologist named Aaron Antonovsky was first to term the condition of salutogenesis meaning the origin of health in 1979. In his theory Aaron was focused into discovering how a person's social class influences their health, and how stress impacts health. (Physiopedia 2021). Salutogenetic experiences on the other hand promote a sense of health. As Aaron said, it promotes a sense of coherence, that is astronauts develop confidence that they will be able to return back home in their safe and sound condition and that all work will go as planned (Lyndon B. Johnson Space Center 2016). This motivates them to keep up to their best work and maintain health in the external and internal environment of the space world. In order to stay positive and healthy in their environment, astronauts are driven by certain factors that keep them going. These may include comprehensibility meaning that they are able to maintain a high and good attitude towards their work and be inspired by it. They should be able to keep good manageability and good control of their emotions, behaviors and desires to fulfill the tasks assigned. Meaningfulness also drives astronauts to see the great value, significance and their purpose of being in space, and also remember the most important things they have left behind on earth in order to maintain a healthy state of mind. The social support the people in space receive from one another and the people they care about really contributes to the ability to stand strong in their rough and unpleasant environment. Spirituality also plays a very significant aspect to astronauts and their happiness, what they believe in and what makes them happy is very important in maintaining salutogenesis in the sense of worshiping and that which brings them joy. Proven by the Patterns in Crew-Initiated Photography of Earth from ISS (Willis, Slack, and Robinson 2006 pg3), salutogenic experiences promote physical well-being. Also the experience of the perception of earth as astronauts view it has become a very important and positive change including the ISS crew members who would be travelling to Mars. The pictures taken by the ISS crew members have also served as an important aspect to scientists and the public.

Suefield (2003) made a divergence between what are called positive environmental aspects and the positive personal and social aspects of space flight. The environmental aspects are identified by two environments being the external and the capsule environment. The external environment consists of mystery (these are parts of the space that scientists find unsolvable), beauty of space and the views of earth. On the other hand the Positive Personal and Social aspect are defined by the astronaut groups which consist of the members of the elite and the superordinate groups. In the same category there are also the post-mission consequences which includes self-confidence, respect, new skills and values. (https://www.physio-pedia.com/salutogenic_approach_to_wellness), (pg. 31). These aspects provided the astronauts with much more faith in surviving and remaining healthy in space. The view of earth in space and the pictures they take to bring them home give them enough leverage to say they still have a lot to look up to hence should hold onto it.

Eudaimonic Well-Being

It is therefore necessary that we carefully look into the well-being of astronauts and cosmonauts who live or spend great time in space. (According to Richard M. and Edward L.2001), the concept of well-being is a very complex construct that has recently been defined in the defense of two terms being the hedonic approach and the eudaimonic approach. The hedonic approach is more focused into what people find to be their happiness, in this term well-being is defined in terms of pleasure and pain avoidance. That is, are people in space able to meet up to their pleasures some of which pleasure my click as the sex drive, listening to one's favorite music and taking a warm bubble bath. Every human being has the things that irritate them differently from one person to another, the things that are unpleasant to them and cause them total disruption and instability. Are they able to avoid them in space, are they able to remain happy in space in the presence of such.

Eudaimonia is the most difficult word to spell and even pronounce, and explains well-being as something that is more than just happiness. It describes happiness in the pretext of the actualization of human potentials. In this platform people in space's happiness depend on whether they have been able to meet their desires, goals, aspirations, likes and dislikes. It is more based on what one individual believes in. The hard work they put into achieving their desired goals becomes their happiness when they finally fulfill them. Many people and almost everyone on earth wishes to see themselves in space, on the moon and/or on Mars, and the stars. The fact is not everyone has the capability of making it on space and for those who do manage to get there it is a dream come true for them, that's what makes them happy and they want to go back to earth and tell everyone what they saw and the whole experience of being there and they sure want to come back to space again. According to the other philosophies (Huta and Waterman, 2014) the concept of eudaimonia was introduced to reflect human flourishing as a concept of virtue and the development of one's full potential. In addition from Aristotle's concept "activity expressing virtue," (NOBASCHOLAR, 2018) eudaimonia refers to that which is worth pursuing in life. Continuing his concept **"defining and operationalizing eudaimonia,"** in the psychological definitions of eudaimonia, describes eudaimonia as a subjective state. It refers to the feelings present when one is moving toward self-realization in terms of developing one's unique individual potentials and furthering their purposes of living, (Waterman, Schwartz, and Conti, 2008 pg42). Eudaimonia still remains an argument from different philosophers including (Waterman 1990), (Ryff and Singer 2008) and (Huta 2015). These three scholars have developed different definitions about eudaimonia as per what they believe what well-being means. For instance, Waterman uses self-realization and personal expressiveness as core definitions features of eudaimonia and suggests that eudaimonia is "activity expressing virtue." On the other hand Ryff defines eudaimonia in a trait-like manner, suggesting that eudaimonia can function very much fully regardless of failure of success and life challenges. Huta treat eudaimonia as a motive to develop the best in oneself. We can therefore try to drive a conclusion that happiness is fully dependent on an individual's capabilities and effort they put into being the best of themselves and finally accomplishing what they worked so hard to get, other than that they can never be happy or have what is called complete well-being. With all that being said it is very evident that we can't really tell the exact definition of what eudaimonia is but we can depend on what individuals believe in.

The argument of what well-being truly is mostly relies on the hedonic and eudaimonic factors and it is a never ending battle between philosophers, religious masters and visionaries. It is therefore important that every individual is able to identify what they believe in and what would make them happy. If being in space is what makes one happy, then they must work hard to be able to have what it takes to be there, and they should embrace the extraordinary time they spend there. It should feel like paradise to finally see what you have been dreaming of becoming a reality, even though there may be challenges accompanying the victory of being in space.

Environmental Challenges and the Psyche (Surviving and Thriving in Space)

The space is a very complicated extraordinary place to be in, considering the environmental setup or its nature of construction. Just setting up from earth, the upside down experience can be very confusing or it is really confusing and fascinating in many ways. Before they go into space, astronauts have to be psychologically prepared in order to be ready for them to be ready to withstand the environmental factors they may be faced with in space, they have to be put in the right psychological frame of mind. Unlike earth, the first thing every individual thinks of is that there is no gravity, so now one can wonder how astronauts are able to spend time in space, how do they walk, how do they eat food, how do they shower, how do they really spend time there and how do the cosmonauts complete the tasks they are assigned to do in space when basically nothing holds them down to the ground. The other thing one can worry and think of is how astronauts and cosmonauts spend and enjoy their free time in space, what activities besides work are they able to do. Can they play some games and if so which ones and how are they able to play them in consideration to the environmental factors. We can therefore continue to speak about the space environment and describe the space environment's six major hazards and their effects on spacecraft. These hazards include charged particles, radiation, atmosphere, free-fall/ gravity, vacuum, and orbital debris. As we all think, space does not have what we normally call free "zero gravity" but rather it has what is called "free fall." Free fall poses three potentially harmful physiological changes to the human body which are decreased hydrostatic gradient (fluid shift) that is there is a change in the way the body distributes its liquids, especially the blood. This may lead to constant urination, decrease in the volume of body plasma, decrease in the production of red blood cells and the heart starts beating faster with greater irregularity. Astronauts also experience altered vestibular function (motion sickness) and reduced load on weight (bearing tissues). (The Space Environment, n.d). Unlike in earth where we have been saved from the radiation and charged particles by the ozone layer, in space the astronauts and cosmonauts are above all the protection hence are more exposed to all the radiation and the galactic cosmic rays. ISS materials bear the brunt of solar proton damage. The Atomic Oxygen bleaches the materials and the Ultraviolet Radiation darkens them. Ultraviolet radiation also damages polymers by either cross hardening or weakening them (NASA ISS Program Science Office, 2005). On the human body, the radiation in space may cause much more serious effects such as cataracts, cancers and death. These effects vary due to the amount of exposure time and intensity, (USAFA ASTRO, 2020). We will now look into how astronauts spend quality time in space and make it an interesting environment for them to survive with less worry about what they have to do. Nowadays space is no longer just about going to Mars, setting up spaceships and installing satellite machines. Space has become more interesting and people want to discover more about it. (Surviving and Thriving in Space, 2021), Dr Chou said there have been several business opportunities created in space that people go there to discover such as space medicine. As one of the people who have been in space he has developed a company called the Explorer Microgravity of which they are anticipating the price of space tourism, they provide services for the travelers before, during and after they return to earth. Such services enable them to monitor if your body will be able to withstand the space environment and for how long it can survive it. Just like on earth, people in space are able to perform several entertainment activities such as playing instruments and producing music, they are able to exercise and keep themselves fit, they are also able to eat food. This helps them to be able to keep their minds off from thinking about their success missions or about making mistakes and creating errors, it helps keep their minds off stress and to refresh reducing anxieties and depression.

Chapter 3

Isolated, Confined and Extreme Environments (ICEs)

Zeest Kadri

Isolated, Confined and Extreme Environments (ICEs)

Although collective efforts go into running successful space expeditions, there is no doubt that the individual astronaut heading into space must be well prepared in order to deal with unexpected situations physically, mentally, and emotionally. Analogue environments to space on Earth are called Isolated, Confined and Extreme Environments (ICEs). ICEs are used to appropriately prepare and train astronauts to effectively venture into space environments under unpredictable conditions.

Long periods of isolation, altered light-dark cycles, confined work environments, and changes in physical environments are some of the challenges that astronauts will have to deal with on their expeditions (Zivi et al., 2020). Additionally, simple conveniences such as using a toilet and bathing that are usually taken for granted will challenge and test an individual venturing into space (Zivi et al., 2020). Other individual changes include stress and confinement; astronauts in space will have to remain away from family and friends for long periods of time, and function with limited communication with loved ones on Earth. The possibility of technological difficulties (ie. through delays in transmission, audio distortion) makes this very socially isolating for an individual (Palinkas, 2001). Furthermore, there may also be other environmental hazards on space ventures that will cause an individual stress- this includes hostile environments, changes in microgravity, exposure to radiation, and micrometeorite collisions (Palinkas, 2001). Despite these challenges, astronauts must be prepared to function autonomously and systematically while having limited communication with Earth (Bartone et al., 2018). As such, an ICEs significantly impacts the outcomes of space missions, in hopes of adequately preparing and training an astronaut for a space mission (Palinkas, 2001).

ICEs on Earth: Is Antarctica an Analogue Environment to Space?

ICEs are somewhat similar to space conditions and are used in various natural and artificial stimulations to prepare incoming astronauts for space expeditions. An individual needs to adapt to these extreme environments, especially with regards to how they affect one's psychological traits, decisions and social interactions. Seeing ways in which humans can adapt and deal with unpredictable ICEs can also streamline research and aid in the development of appropriate technologies and protocols that can contribute to adaptation in space environments and result in successful space missions. Polar settings on Earth, notably Antarctica, are popular locations hailed as ICEs.

Antarctica is hailed as one of the most reliable space analogues, and can help to prepare for various challenges that one will definitely face on space trips. In particular, it is a region where the effects of isolation, confinement, environmental hazards and changes in light-day cycles can be studied to effectively prepare individuals to head into space successfully.

Located at the Antarctic polar circle at extremely high latitudes, Antarctica is a place that is covered in about 14-million square miles of ice layers (Tortello et al., 2018). The region is characterized by ice sheets, glaciers and sea ice (Australian Antarctic Program, 2020). Ice sheets can be up to 4km thick, and beneath the ice layers are geological formations of various mountains and valleys (Australian Antarctic Program, 2020). There are also limited plants and creatures, some of which include penguins and various forms of flora and algae (Australian Antarctic Program, 2020). There also may be life in unexpected places underneath sheets of ice in Antarctica. Some individuals think that this possibility of hidden and buried lakes is analogous to the possibility of discovering and understanding the geology of icy moons in space such as Europa (Jupiter's moon) and Enceladus (Saturn's moon) (Cool Antarctica, n.d.).

Some extremely hazardous environmental conditions and phenomena that occur in this region include low visibility, heavy winds, tsunamis, landslides, blowing snow, volcanic eruptions, blizzards and avalanches (Australian Antarctic Program, 2020). This region is also the windiest on Earth; low pressure wind systems (called the circumpolar trough) with interior high pressures result in katabatic winds, which produce cold and dense air that accelerate downwards at very high speeds over large areas (Australian Antarctic Program, 2020). Wind speeds are usually more than 100km/hr but have been measured as being as fast as 200km/hr. Additionally, these wind speeds can last for more than a few days at a time (Australian Antarctic Program, 2020). The Antarctic region is also very dry, with little precipitation. Temperatures around the coast are about -10C and as low as -60C in interior parts. In the summer, temperatures can get as high as 10C (near the coast) in the summer and -30 inland (Australian Antarctic Program, 2020). The lowest temperature ever recorded was about -90C (Australian Antarctic Program, 2020). Another phenomenon that is faced by individuals venturing to Antarctica for special training/scientific expeditions is Whiteout. This is an optical phenomenon that makes it extremely hard to distinguish direction and orientation; one may not be able to distinguish even their shadows, certain landmarks or the horizon because of unbroken snow and overcast skies (Australian Antarctic Program, 2020). These hazardous environmental conditions necessitate that individuals wear appropriate clothing and equipment in both outer space and Antarctica to carry out their respective duties and missions. In outer space, some environmental hazards that need to constantly be dealt with include micrometeorites, vacuums, radiation (both solar and ionizing), surface charging/arcing, temperature extremes, thermal cycling, micrometeoroids, orbital debris as well as environment induced contamination (National Aeronautics and Space Administration [NASA], n.d.). Although these challenges are not exactly the same so to say, an analogue ICE environment suggests that an individual will be more prepared to effectively deal with various dangerous and unpredictable conditions.

Aside from the hazardous environmental conditions, the tilt of the Earth results in only two seasons in the region. There are 6 consecutive months of daylight and 6 consecutive months of darkness; the polar night is characterized by 24 hours of darkness during the wintertime and constant light conditions during the daytime (Tortello et al., 2018). This can seriously impact an individual's circadian rhythms and results in changes in physiological markers and hormones. The changes in these light cycles impact an individual's circadian rhythms and can result in seasonal affective disorder (Tortello et al., 2018). In space and Antarctica, the absence of natural light exposure (depending on time and location) can impact an astronaut's mental health. Research shows that almost five percent of individuals meet DSM/ICD criteria for certain psychiatric conditions such as anxiety and depression (Tortello et al., 2018). Furthermore, poor sleeping conditions and other forms of psychosocial stress are faced by both space and Antarctica, which is another similar element that reveals how Antarctica is a very similar analogue to outer space.

The use of satellites are vital with Antarctic expeditions as they are used for communication, surveillance, navigation, intelligence-gathering, search and rescues, fishing and research (Byers, 2020). Similarly, many space missions rely on satellite technologies that help them to transmit data and other information from space to the Earth and vice versa (LaFrance, 2015). Today, as interest in space exploration continues to grow, it is important to realize that one region or country does not necessarily have any economic or territorial claims to outer space (Salazar, 2017). This case also applies to Antarctica, which demonstrates that international cooperation is facilitated in both regions (Salazar, 2017).

Many of the aforementioned characteristics make Antarctica a dangerous but equally mysterious space, just like outer space itself. Aside from the lack of gravity, many similar challenges are faced by individuals in both environments.This includes terrestrial conditions and other logistical and technological methods used in both regions. The lack of access to supplies that sustain life, isolation, hazardous weather/environmental conditions, and abnormal daylight cycles make Antarctica an environment that can be considered an analogue to space.

Isolation and Confinement: Effects and Ways to Overcome it

One of the most strikingly similar conditions that potential astronauts will experience in both Antarctica and space is prolonged isolation and confinement. Isolation and confinement are vital characteristics that are studied in both Antarctic and space settings as it is possible that they impact one's problem solving skills, group dynamics and overall mood (Golden et al., 2018). They can play a role in one's overall resilience and growth (Golden et al., 2018). Over the past few years, research efforts have been focused to better understand the various psychological and pathophysiological impacts of both the analogue ICE and space environment. This includes research that works to understand the mental and emotional states of being exposed to these conditions for a prolonged period of time (Dunbar, 2007; Golden et al., 2018).

ICEs environments suggest that psychosocial elements of behavior and performance are likely to have a significant impact on the outcome of long-duration missions in space. ICE environments impact the psychosocial elements of behaviour and impact not just an individual's performance, but their overall long-term mission goals and duties (Palinkas, 2001). Even elaborate and modern screening techniques do not guarantee that an individual may/may not develop psychological problems over their isolated journeys. This is because missions to both Antarctica and outer space consist of limited communication from loved ones and friends back at home, demanding mission goals, and hazardous surroundings (Palinkas, 2001; Temp et al., 2020). Other stressors that arise include boredom, emotional deprivation and reduced opportunity to avoid interpersonal conflict (Temp et al., 2020). Research has found that a single crew member suffering from mental or emotional distress can contribute to safety issues (Pagel & Choukèr, 2016). The effects of isolation and confinement may result in a lack of vigilance which can result in potential damage to station buildings or result in careless behaviours (Pagel & Choukèr, 2016).

One gap in research is that the research comparing ICE environments to outer space is very limited and outdated. Fortunately, innovative technologies and strategies today have paved the way in order to help individuals better deal with the issues of isolation and confinement. Missions to both Antarctica and outer space usually last for many months. NASA outlines some ways that astronauts have recently been using to overcome feelings of isolation on their ICE missions to space in order to improve their health and wellbeing. They have created an acronym called CONNECT, which stands for Community, Openness, Networking, Needs, Expeditionary Mindset, Countermeasures, Training (Perez, 2020).

First, astronauts who deal with isolation should know that their work has an exceptionnary impact on both their local and international "community", giving an individual a sense of purpose and motivation to their actions and decisions (Perez, 2020). Next, Openness is a trait that should be adopted in order to appropriately adapt to life's changes. Although astronauts venture to space in small groups, "Networking" through technological functions such as video chat (ie. to loved ones or professionals like psychologists) is encouraged in order to help individuals reduce their feelings of isolation (Perez, 2020). Care packages and virtual assistants are other ways in which networking can be somewhat maintained in these environments. Next, an individual's "Needs" should be considered - they need to be fit in a physical and emotional sense, which can be done by following a strict exercise, eating, leisure working and sleep schedules (Perez, 2020). Furthermore there are technologies which exist that can alter the lighting present in various ICEs which can improve the duration and quality of sleep of astronauts. An "Exceptionary Mindset" can be facilitated through respecting different viewpoints and experiences of other individuals, and by promoting harmony in these environments through appropriate conflict and support management (Perez, 2020). Next, "Countermeasures" aim to reduce the stress of isolation, and it is encouraged that individuals use coping strategies (via mindfulness or journal writing) in order to counteract the stresses of isolation while in space (Perez, 2020). Finally, appropriate "Training and preparation" may arguably be the most important aspect to maintain an individual's wellbeing while in ICEs, individual crew members in space will be training in analogue environments in order to ensure that they are adequately prepared in the best way possible to manage the challenges that come with space exploration (Perez, 2020).

Overall, it can definitely be seen that isolation and confinement is a significant issue that individuals face when venturing into ICEs. They not only impact the individual, but the overall mission as a whole. As increased research and subsequent innovative technologies continue to develop, there is no doubt that individuals venturing into ICE environments such as Antarctica and outer space will face significantly fewer impacts of isolation and confinement.

Conclusion

Antarctica is an analogue ICE to outer space, and individuals in both locations have to deal with many challenges and unpredictable situations. An Antarctic setting, despite being completely different from space, serves as a location where prospective astronauts can effectively prepare and train to successfully succeed in their respective space ventures. Both locations cause extreme isolation and confinement, and also expose individuals to extreme environmental hazards which can impact both their physical and mental health. Today, various innovative strategies have been proposed by organizations such as NASA to reduce the impacts of isolation and confinement in ICEs.

Chapter 4

Sleep and space psychology

Maya Nagorski

Human physiology undergoes many subtle changes during a cycle that is approximately 24 hours, including secretion levels of certain hormones, body temperature, and body system activity (Germain & Kupfer, 2008). This cycle, known as the circadian rhythm, is controlled by the hypothalamus and largely kept in sync by a person's exposure to light (Guo et al., 2014). Irregular sleep schedules, disruptions of the natural light-dark cycle due to artificial light, or jet lag can all interfere with one's circadian rhythms and the equilibrium of their body. Once a person has become entrenched in an environment without these natural light cues, their circadian rhythms are referred to as "free-running", as their circadian period begins to shorten (Sack et al., 2007). The effects this has on a person's health, both mental and physical, can be tremendous, even possibly contributing to the onset of neurodegenerative diseases (Wulff et al., 2010). Furthermore, a disturbed circadian rhythm is common in patients suffering from depression or other mood disorders (Germain & Kupfer, 2008). While the onset of such mental illnesses would be devastating to anyone in any profession, there are few environments where such effects would be as disruptive as a spacecraft. The physical and mental health of astronauts are essential to their performance, and while they may extensively train their bodies, it is more difficult to prevent or treat something as unpredictable as irregular sleep.

Observations by Kanas et al. (2008), Frost et al. (1976), and Santy et al. (1988) have found that, due to the nature of space travel, astronauts sleep much less than they do on Earth, with several crew members reporting having slept as little as under 5 hours a day on average (Wu et al., 2018). This level of sleep deprivation may only be found in extreme cases, as a report done by NASA concluded that astronauts complete 6 hours of sleep per day on average. However, for the average age of astronauts, 34 (Blodgett, 2018), this is still under the NASA recommended time of 8 hours (May, 2015). Moreover, NASA has reported that a staggering 45% of astronauts, over a course of 79 missions, decided to use medication to aid them in sleeping (Wu et al., 2018). Unfortunately, popular medications such as melatonin have often been observed to be ineffective, perhaps due to the unique motion of the spacecraft or other environmental factors (Guo et al., 2014).

There are several factors that could affect the natural sleep schedules of astronauts, and therefore their circadian rhythms. Astronauts may experience difficulty sleeping due to the stress of their environment, the effects of microgravity, and the lack of a natural day-night cycle (Mallis & DeRoshia, 2005). The natural light they receive from outside the spacecraft can be extremely confusing for their bodies; while orbiting the Earth, an astronaut will observe a sunrise or sunset every 90 minutes (Czeisler et al., 2002). The artificial light on space stations, even if adjusted to mimic a 24 hour cycle, is not sufficient enough to keep one's circadian clock in sync, due to its low intensity (Guo et al., 2014.) Another factor in the disturbed sleep of an astronaut is the typical demand for shift work. Crewmembers often sleep in rotation, in order to always have someone on hand (Mallis & DeRoshia, 2005). This type of scheduling could have any given astronaut at odds with their habitual sleeping-waking cycle, and often leads to a decrease in performance quality, as it did in the Salyut-6 mission.

Gundel et al. (1993) investigated the physiological effects of such conditions on astronauts undergoing short missions orbiting the Earth. During this study they monitored length of sleep, REM cycles, mood, body temperature, and alertness of a single astronaut working in the China Astronaut Research and Training Center. Results indicated that the amount of time between falling asleep and entering the REM state, also known as REM latency, as well as the second REM cycle itself was significantly shortened. Shortened REM latency has been found to be a common characteristic in depressed patients, and is often used as a diagnostic tool (Kupfer, 1976). Interestingly, they noted that while on Earth, this subject's alertness typically dipped before sleep (Gundel et al., 1993). However, on this space mission, alertness instead dipped after sleep, indicating a delay in his physiological cycles. Additionally, it was noted that the participant's slow-wave activity changed, which falls in line with a model detailed by Wu et al. (2018) wherein homeostasis is regulated during sleep in part by the "S process". Disruption of this process can be noted through variations in slow-wave activity. Therefore, though it was impossible to determine in this 8 day period whether or not the subject's circadian rhythms had become free-running, there was significant disruption to his physiology.

In terms of extended missions on a spacecraft, decreased mental performance due to lack of sleep and disrupted circadian rhythms can be significant. In general, sleep deprivation can negatively affect mood, memory, reaction time, and behavioural inhibition (Orzeł-Gryglewska, 2010). However, a study was conducted in NASA's Human Exploration Research Analog in order to observe the performance of sleep deprived subjects specifically in an environment similar to that of a spacecraft (Nasrini et al., 2020). While experiencing confinement, isolation, simulated tasks, and simulated stressors, subjects were either permitted to sleep in shortened intervals, or completely deprived of a night's rest. Subjects allowed a shortened rest period showed a significantly decreased ability to correctly recognize emotions in others. While this effect was not seen in the subjects completely deprived of sleep, these subjects showed a significant decrease in speed and accuracy during attention, sensory-motor, and visual-tracking exercises.

These effects can be disastrous on a spacecraft, for a number of reasons. Firstly, much of an astronaut's workload is composed of maintenance, which requires both attention and dexterity (Guo et al., 2014). Furthermore, crewmembers are required to exercise each day, for long stretches of time (Government of Canada, 2019). This must be done in order to combat the detrimental effects that weightlessness has on the human body; without this routine, astronauts risk bone deterioration, weakening muscles, or a weakening heart. The fatigue and decrease in motor control caused by sleep deprivation is therefore a great danger. Finally, one of the greatest challenges in space travel is the mental strain of isolation. Cooperation and camaraderie within the crew is extremely important, and can easily be damaged by the imbalances in mood prevalent in the sleep deprived. As discussed by Palinkas (2007), social stress can be a determining factor in the success of a given trip. As noted in this same study, the crew dynamic is already likely to deteriorate over the course of the mission; added animosity due exhaustion can only hurt.

With these serious side-effects and health hazards in mind, many space travel programs go to great lengths to either prevent or treat sleep deprivation and circadian rhythm disruption. NASA especially has come to emphasize the importance of education, bringing in researchers to investigate the effects of fatigue and propose solutions. (Mallis & DeRoshia, 2005) Further, when preparing for the Mars Exploration Rover mission, they held educational workshops to train crewmembers in dealing with sleeplessness and its effects. In terms of actionable changes made on space crafts, several steps have been taken. For example, as mentioned before, the lights found on spacecraft are insufficient in mimicking the natural day-night cycle, due to their low brightness. To properly simulate light and darkness, astronauts may cycle between using bright lights or dark goggles (Guo et al., 2014). Light treatment appears to be quite effective, as in previous experiments, researchers were able to shift the cycle of participants' secretion of hormones such as melatonin or cortisol (Mallis & DeRoshia, 2005). The most frequently employed method of preserving an astronaut's natural physiological cycles is simply to properly schedule their shifts in order to account for stress, work demands, and physical necessities such as exercise. Minimizing interruptions or the need for shift work immediately eliminates some of the greatest challenges to providing a proper period of rest. There is an effort to afford astronauts as consisten a sleep schedule as possible, with guidelines prohibiting the significant lengthening or shortening of their rest unless it is absolutely necessary (Mallis & DeRoshia, 2005). Research has been done into how to minimize the detrimental effects if disturbed or shortened rest periods are absolutely required. Banks et al. (2010) found that if participants were granted extra sleep, in order to recover from a shortened sleep period, the effect on behaviours such as attention, alertness, and visual recognition was not so drastic. Furthermore, as discussed by Buguet (2007), sleep following a period of exercise can be extremely restful, due to familiar physical strain.

While space travel programs across the world take this issue very seriously, it is interesting to examine how different countries have tackled the problem, and how astronauts in different countries report on the quality of their sleep. For example, China has dedicated significant effort towards researching sleep problems and their solutions (Wu et al., 2018). One possibility they have considered is allowing each individual astronaut to determine their sleep routine according to their own needs, as opposed to using a predetermined schedule.. There have also been attempts to incorporate Tai Chi into the astronauts routines, in order to help with sleep, though this has not been properly researched or experimented. Russia, similarly, has conducted research into how to combat this problem, such as their observation of both short and long-term missions between 2006 and 2011 (Barger et al., 2014). This extremely comprehensive study recorded the sleeping and waking of astronauts in the time leading up to the mission, the mission itself, and a short period after the mission. They gave specific consideration to the frequency of sleeping pill use, and whether or not the continuous use of such medication was feasible, due to the possibility of dependence or mental impediment.

While work has been done to remedy this problem, further research is necessary for ensuring the health of crew members during space travel. To begin with, new medication must be designed, specifically created to treat sleep disorders in space (Guo et al., 2014). The lackluster effect melatonin had on the STS-90 mission, along with the previously discussed dependence astronauts have on sleep medication, proves that a new drug must be created with the environmental factors of space in mind. Furthermore, it would be prudent to conduct studies into preparatory training on Earth, in order to reduce fatigue. Is it possible to properly sync an astronaut's waking-sleeping cycle to what they will be experiencing in space, in order to reduce stress during the mission? As Buguet (2007) discussed, it is easier to access sleep with the aid of previously established neural pathways. Would it be worthwhile to train astronauts to fall asleep easier using routines, such as always performing a certain task before sleeping? Another possibly advantageous avenue of research is looking into creating a specific environment on spacecrafts conducive to sleep. Three of the most frequent complaints from crewmembers, in terms of obstacles to sleep, are noise, uncomfortable temperatures and motion sickness (Wu et al., 2018). Efforts have been made by NASA in the past to create comfortable sleep stations, but to little effect (Barger et al., 2014). An observational study was carried out on the STS-50 in 1993 in order to examine how effectively sleep stations could block out sound: all crewmembers reported being woken several times a night (Koros et al., 1993). Finally, further research must be conducted into the long term effects of sleep deprivation and desynchronized circadian rhythms, particularly in the cases of longer missions. What unique challenges might an astronaut face in rehabilitating themselves to Earth, specifically in regards to mood disorders or other mental illnesses? What can be expected from the recovery period, in terms of both length and difficulty? What programs or treatments may be effective in aiding them? All of these questions, and more, would be useful in minimizing and addressing the effects of disordered sleep in space.

Circadian rhythms and their maintenance through proper sleep are paramount to ensuring the health of any given person; this only becomes more important in the context of space travel. Astronauts often report distrubed or shortened periods of rest, resulting in desynchronized physiological processes and decreased performance. In such a high pressure environment, a lacking mental or physical state can prove to be disastrous, both for the success of the mission and the long term health of its participants. Several measures have been put in place by programs such as NASA, as well as research efforts into improving the space travel experience. Other countries, such as China and Russia, have made similar attempts to understand and treat sleep deprivation. With this in mind, there is still innovation to be reached, and further questions to answer. Putting humans in an environment as dangerous and unfamiliar as space will always come with risks; this is rarely highlighted as clearly as when considering the effect space travel has on the extremely important process of sleep. However, with appropriate research, understanding, and countermeasures, it is possible to minimize these risks; this has never seemed as feasible as it does in the present.

Chapter 5

Mood disorders, neurasthenia, psychosomatic reactions

Kirithika Bharatselvam

Space flight constitutes a massive change in a person's normal day-to-day life, inevitably causing various mental and psychological issues to stem from this huge change of environment. Factors such as separation from friends/relatives, long periods of isolation, feelings of loneliness, and other stressors (Oluwafemi et al., 2021) contribute to an astronaut's risk of developing mood disorders like depression and anxiety, neurasthenia and adjustment reactions.

Transient Anxiety and Depression in Space

Psychiatric issues like depression and anxiety occur as a result of stressors that astronauts face during space missions. A stressor can be defined as an environmental stimulus that negatively affects a person. Habitability, physical, interpersonal, and psychological stressors are the four stressors that can be experienced in space (Kanas & Manzey, 2008, p. 1). The first stressor, habitability, includes examples like the air quality, temperature, lighting, presence of vibrations, ambient noises, etc. The second stressor, physical stressors, includes light/dark cycles, acceleration, microgravity, etc. Psychological stressors are the third stressor consisting of isolation, confinement, danger, workload, etc. And finally, interpersonal stressors include gender issues, leadership issues, the size of crew, cultural differences, etc (Kanas & Manzey, 2008, p. 2). These stressors can cause psychiatric stresses like depression, anxiety, asthenia, and adjustment disorders amongst the space crew. Some stressors listed above can also be related to other stressors from the same category or another (Kanas & Manzey, 2008, p. 1).

Psychological stressors like isolation, confinement and risk of danger are found to be stressors that mostly cause anxiety and depression, amongst other psychological issues. As there are very few space mission studies due to their expensive and political nature, many other space-related studies conducted on environments on earth can give insight into psychiatric stresses that occur during human space missions (Salamon et al., 2018). The crafts used during early space exploration programs can be classified as ICE environments (isolated and confined environment), as inhabitants of the craft are separated physically from (conventional) support systems and other people in a confined space (Salamon et al., 2018). Studies from past spaceflights and other ICE analogue studies have shown behavioural problems like anxiety, depression, anger, interpersonal conflict, sleep deprivation, social withdrawal, decrement in motivation and in-group cohesion from occurring due to the psycho-environmental factors of ICE habitats. These factors include sensory restriction, social isolation, crowding, and lack of privacy. Behavioural issues can be exacerbated when the isolation period is extended during long space missions (Salamon et al., 2018). As missions become longer over time, like the missions to Mars, it heightens the feelings of isolation and loneliness amongst crewmembers compared to past space missions of shorter duration periods. Long space missions can also potentially cause emotional deterioration and conflict, creating more stress and other problematic behaviours, which ultimately interferes with the crewmembers' relationship with each other and their own productivity (Salamon et al., 2018). Astronauts can also become homesick during long-duration missions, increasing their risk of developing depression and anxiety amongst other issues, including absent-mindedness, phobic avoidance, withdrawal, and the tendency to hyper-fixate on their home on earth (Oluwafemi et al., 2021). Overall, this can jeopardize the effectiveness of a space mission and must be addressed in order for a successful mission to be completed.

Emergencies are a possible occurrence faced by astronauts during spaceflight due to crew action or naturally. When an emergency happens, it can put immense social and psychological stress on an individual. Disasters-affected persons are highly likely to experience heightened levels of distress and fear, and these psychological reactions can be either brief (wherein mild cases psychological support is unneeded and people can rebound back to normal) or develop into psychiatric disorders (Oluwafemi et al., 2021). This happens when an individual encounters seemingly life-threatening situations and can develop post-traumatic stress disorder (PTSD) amidst other psychiatric issues. Other disorders like generalized anxiety disorder (GAD) and major depressive disorder (MDD) can be comorbid with PTSD and are commonly seen amongst disaster survivors (Oluwafemi et al., 2021). Psychiatric conditions can hinder an astronaut's ability to function which is needed to be at a high level for a space mission to be completed and is essential to address to prevent this from occurring.

A U.S. study on five Antarctic stations conducted by Gunderson (1968) shows the psychological/emotional issues occurring at these stations before and after the Wintering-over period (Kanas & Manzey, 2008, p. 95; PhD, 1968). These stations were excellent space simulation environments on earth due to their degree of isolation, danger and size. Gunderson found a higher incidence of psychological symptoms amongst naval servicemen than in the general population, which consisted mainly of technicians and scientists. During the wintering-over period, rates of depression increased by 15% among naval personnel, and they overall experienced more psychological problems than the civilians even though the naval personnel were used to being active and outside (Kanas & Manzey, 2008, p. 95).

Another stressor that primarily causes anxiety and depression, amongst other psychological issues, are interpersonal stressors (Kanas & Manzey, 2008, p. 2). Interpersonal stressors like cultural differences can create problems amongst crewmates due to characteristics like how one expresses their emotions is common in some cultures while being unfamiliar in others. In addition, psychological issues can occur differently amongst cultural groups . For instance, depressed mood comorbid with fatigue for Russians, however for Americans, it comorbid with anxiety. Decision-making and individual behaviour also differ amongst cultures, for example, grooming habits and the need for privacy varying depending on the situation. Furthermore, social behaviour such as whether it is appropriate to socialize during meals or how hosts are expected to treat their guests are all social behaviours that can create tension and manifest into psychological issues (Kanas et al., 2009). These psychological difficulties can affect both the individual and the entire crew. For instance, an astronaut who develops anxiety may not be able to complete tasks perfectly, especially during an emergency which then creates an uneasy atmosphere for fellow crew members.

Cultural differences and their impact on an astronaut developing mood disorders can be seen in missions to the International Space Station (ISS) and the Mir space station. The ISS sample study consisted of more American members, including eight astronauts and 108 mission control personnel, and only consisted of twenty Russian mission control personnel and nine Russian cosmonauts (Kanas et al., 2009). Because the number of Americans outnumbered the Russians, anxiety and tension likely rose as a result of cultural differences and organizational culture. For instance, when issues arise, the Russian space program generally relies on gaining opinions from experts rather than using written procedures (Kanas et al., 2009). Therefore, this contrast (among other differences) could create an uncomfortable environment for the Russians that may add to the development of psychiatric disorders.

Asthenization and Adjustment Reactions

Asthenia is a disorder that, according to Russian flight surgeons and space psychologists, is a critical condition that can affect individuals in space (Kanas et al., 2009). Asthenia can be defined as an excessive response to a stressful event or changes in a person's life (Kanas & Manzey, 2008; Sandoval et al., 2021). Asthenia also referred to as "asthenization" (Kanas & Manzey, 2008, p. 146) is a syndrome that is often defined as weakening of the nervous system, resulting in irritability, restlessness, decreased performance and concentration, fatigue, exhaustion, sleep disturbances, physical weakness, and decrements in motivation and appetite (Kanas et al., 2009; Salamon et al., 2018). This may entail states of depression, euphoria and emphasizing ones' negative personality traits (Kanas & Manzey, 2008, p. 37). This syndrome (Asthenia) is observed as a milder variation of neurasthenia, a severe mental disorder that appears in the ICD-10 system of classification and requires treatment (Kanas & Manzey, 2008, p. 144).

Roots of neurasthenia come from George Beard, an American physician. Many of his patients expressed symptoms of hopelessness, exhaustion, mental irritability, morbid fears and anxieties, difficulties in concentration, insomnia, bad dreams, forgetfulness, pains, sexual issues, and headaches. Beard thought that these symptoms resulted from an underlying physiological disorder that he indicated as nervous exhaustion or neurasthenia (Kanas & Manzey, 2008, p. 145). He observed neurasthenia as a disease that mainly affected upper-class Americans. Though, both neurasthenia and asthenia are not diagnostic entities in the American Psychiatric Association DSM-IV (Kanas & Manzey, 2008, p. 145). Thus this syndrome cannot be diagnosed in patients by American mental health using this manual. However, numerous asthenia symptoms are included under diagnoses such as adjustment, dysthymic, chronic fatigue syndrome, or major depressive disorders in America (Kanas & Manzey, 2008, p. 145).

Asthenia is seen as a major issue affecting most cosmonauts' emotional states in long-term space missions according to the Russian space program (Sandoval et al., 2021, p. 21). Additionally, it has only been seen by Russian medical psychological and medical personnel in cosmonauts after spending time living in space for four months (Kanas & Manzey, 2008, p. 146). However, neurasthenic spectrum disorders also show in the international diagnostic system used in Russia, China, and Europe (Kanas et al., 2009). Asthenia in the mid-1970s was considered a syndrome that occurred as an adaptive reaction because of the nervous system becoming exhausted due to overexertion, intracellular metabolism disruptions, lack of proper nutrition, and intoxication (Sandoval et al., 2021, p. 21).

Later on, Aleksandrovskiy stated that in space, asthenia develops in three phases, stage one (hyperesthesia), where sensitivity is generally increased by external stimuli that result in increased (sometimes pointless) activity and hyper-arousal, emotional irritability and instability, impatience, poor attention and concentration, decreased memory, headaches, sweating, fatigue, sleep disturbances, and instability of blood pressure and the pulse. In stage two, an individual experiences irritability, irritable weakness, and emotional instability that progresses into severe fatigue, somnolence, and negative emotional reactions. Finally, in the third stage, there is indifference and inertness, fatigue is constant, passiveness, apathy, and lack of work capability (Sandoval et al., 2021, pp. 21, 22). However, Myasnikov also concluded that asthenia syndrome is identified in three stages in space. The first stage is defined by heightened emotional excitability, the second stage is expressed by symptoms such as frequent fatigue, decrease in performance quality, mood swings, and signs of sleep disturbance. And the third stage is characterized by low mood, conflict tension, expressed irritability, frequent and significant errors in performing work-related tasks, hypochondriac phenomena, and sleep disturbances that require the use of sleep aid medication (Sandoval et al., 2021, p. 22).

Symptoms of asthenia in space as described by Myasnikov & Zamaletdinov, as being similar to those affected on the ground, and both involve de-adaptation to a stress-inducing environment (Sandoval et al., 2021, p. 22). Though Myasnikov prefers using the term "psychic asthenization" when referring to the syndrome in space (Kanas & Manzey, 2008, p. 146). On the contrary, researchers from the United States specifically used the Profile of Mood States (POMS) to look at asthenia in space. The presence of asthenia in space was not demonstrated using the POMS due to the POMS not addressing physiological aspects of the syndrome but only the emotional aspects (Sandoval et al., 2021, p. 22). However, signs and symptoms that may suggest asthenia have been anecdotally reported by American astronauts during long-term space missions (Kanas & Manzey, 2008, p. 144).Overall, asthenia is not universally classified as a disorder, but cosmonauts frequently report it, suggesting that an astronaut's cultural background can impact how they handle ICE environments (Salamon et al., 2018). Adding on, it may be an issue if flight surgeons only utilize the U.S. system as it may potentially under-detect asthenia among crewmembers. Though, there is evidence that mood state patterns systematically vary amongst cosmonauts versus astronauts (Kanas et al., 2009).

Asthenia can cause errors in operation, conflicts between crewmembers, and impair performance . The condition is due to an accumulation of fatigue that has grown over time. Russian experts monitor and diagnose asthenia in space by analyzing medical information, conversations between crewmembers and personnel in mission control while also clinically scaling somatic symptoms, mood, fatigue, and sleep quality (Kanas & Manzey, 2008, p. 146). Asthenia in the United States has symptoms that can be classified under adjustment reactions (Kanas & Manzey, 2008, p. 145), which is one of the most reported psychiatric problems in space (Kanas et al., 2009). It can be defined as an unusual response to stressors that are either internal or external. For instance, at the start of a long-duration space mission, one astronaut gained symptoms of depression due to feelings of isolation on-orbit in addition to being separated from his family (Kanas et al., 2009).

Conclusion

Overall, it is vital to address the possibilities of astronauts and cosmonauts being diagnosed with psychiatric disorders like anxiety, depression, asthenia, and other adjustments to provide astronauts with the best care and prevent an unsuccessful mission from occurring. Issues such as anxiety, depression, feelings of uneasiness, interpersonal conflict, sleep deprivation, social withdrawal, etc, can result in reduced work effort and increased tension among the crew, possibly putting the space mission at risk. Identifying the possible stressors that can result in these psychiatric issues, such as psychological stressors like isolation, confinement and risk of danger and Interpersonal stressors like cultural differences, can help prevent psychiatric problems from developing in astronauts in the future. Addressing and understanding the reasons for asthenia, adjustment disorders, depression, and anxiety, among other psychological issues, can help mitigate these issues by prioritizing different countermeasures to ensure a more comfortable environment for astronauts and cosmonauts alike.

Chapter 6

Astronaut Screening

Zain Kadri

Astronaut screening, also known as the Astronaut Aptitude test, contains a numerous amount of questions and psychological tests in respects to the official NASA astronaut candidate requirements. These tests are taken to determine if you are capable of space exploration. 15 questions are included in the astronaut screening, you are judged based off of 4 essential criteria:

• Physicality and medical conditions
• Spatial visualization
• Knowledge, education and Abstract thinking (IQ)
• Personality

(CSA). Astronaut screening is made to determine whether or not an individual is qualified for the job. It is important to be proficient and all rounded in the listed fields above before applying for a job as an astronaut (NASA). Astronaut screening is a mandatory test made to test applicants beyond their limits. This gives the company a rough idea of how you work and gives them a good idea of how you will work around others collaboratively. Generally speaking, this will let the company know what work they can expect of you (CSA).

Space companies generally look for applicants during their primal age, ranging from ages 26-46 to ensure good quality work. Younger applicants are preferred compared to older ones, as the younger ones will be able to work for longer. Many requirements are needed to be reached and followed before being accepted as an astronaut. Below you can find a general idea of the requirements you need to reach to become an astronaut.

Physicality & Health

Physically, astronauts should be fit according to a certain standard. Astronauts are strictly required to have good health and follow the given criteria. The touchstone of your height must be 4'9, and the max should be 6'2. Your weight must be within the limit of 210lbs and not less than 110lbs (Matsumo). Astronauts are obligated to have 20/20 vision (with glasses) and should not be colour blind. Future astronauts should be able to hear perfectly, with no chronic or recurring auditory circumstances. Finally, astronauts must not have blood pressure over 140/90mm in a sitting position. (CSA)

All candidates must pass the European part MED class 2 examination by a professional and certified medical examiner (CSA). All applicants with disabilities must have a physician to be able to give a medical certificate stating that without their underlying condition, they are able to abide the European Part MED class 2 examination (NASA). Applicants must be free of any disease, addiction or dependency on drugs, tobacco or alcohol. Fluidity and joint movement should be normal (specific disabilities are exceptions), and no psychiatric disorders should be present either. (NASA)

Overall fitness is also a big requirement when completing an astronaut screening. Astronaut screening fitness can be challenging, as they push you to your limits with swimming tests, physicality tests, cardio tests and weight tests (Pubmed). During fitness testing time, the fittest is not the one who gets chosen, it's the one who has a balance in all areas (OCHMO).

Undoubtedly, these health standards and expectations are high, but have to be reached in order to have a successful space mission (IOM). Most applicants do not make it past these tests as they are not physically fit, or have a dependence on alcohol and tobacco (CSA).

Spatial Visualization

In an astronaut screening, spatial visualization is crucial in order to mentally comprehend our environment and react accordingly. Applicants must have 20/20 vision (with glasses), and pictures of their eyeballs from front and back are taken as well as an ultrasound of the optic nerve to study changes within the eye (Pubmed). Applicants are also tested on their hand-eye coordination by balancing on one foot while trying to throw a tennis ball in a designated area or target. Another 30 second agility test is performed to assess hand-eye coordination further, this is by testing how long a series of movements can be completed within an allotted time.

Spatial visualization is a key component to being an astronaut, laser surgery as well as glasses are accepted when testing for the job. In fact, 80% of astronauts and pilots use glasses and it is quite normal to have (HIDH). After receiving a laser surgery for the eye, testing is only permitted 1 year post-surgery (Pubmed).

Knowledge, Education & Abstract Thinking (IQ)

To be accepted as an applicant, it is vital that you have at least a bachelor's degree in the field of engineering or science, you may also have a doctorate in the field of dentistry or medicine (NASA). On top of those requirements, candidates are asked to have 3 years of professional experience in a related field. Applicants must be able to communicate efficiently by getting their point across with verbal and oral communication. Within the CSA, highly proficient French and English are a requirement as an applicant. Job seekers can be called in at random to provide certifications and perform technical work to further test applicants (NASA).

After checking all of the previous boxes, it is vital to have abstract thinking. An IQ test is taken to further test your intelligence and knowledge. At NASA, a score of 130 or higher is needed to become a full time astronaut. Furthermore, collaborative thinking and communication while working on a given task is an absolute mandatory skill in order to provide the best results in a space mission. Applicants will be put in groups and given certain tasks and projects to complete and will be tested based on their communication, knowledge and understanding.

In some screenings, applicants must solve simple problems and puzzles underwater in an 8 ft pool. Applicants must be well-rounded in their knowledge to reserve a spot in the CSA or NASA.

Personality

During an astronaut screening, they assess your personality heavily over all other components. The CSA, as well as NASA look for individuals with initiative, leading and investigative personality traits. This is important when finalizing a candidate to settle on as this will drive the basis of a successful space mission. They also look for applicants with a positive mindset, this is because compared to someone with a negative mindset, you are more likely to succeed at something. Personality is extremely important when determining a successful candidate to provide other chosen candidates with a good environment to work in (Johnston).

Self Management

Undoubtedly, astronauts undergo heavy amounts of stress as they progress in their career. Managing your time, schedule, tasks and social life all at once is hard. Astronauts must be disciplined and sacrifice free time in order to balance their schedules. Many astronauts lose sleep over the extremely busy schedules and strict work environments they are put in. As time goes on, astronauts become used to their schedule, however they still undergo large amounts of stress.

One way astronauts deal with stress is by creating a timed schedule, which they would follow in order to complete the maximum amount of work throughout a day. Astronauts complete their work by using the pomodoro technique. The pomodoro technique created by Francesco Cirillo is where you divide your work in different parts, work for 30 minutes at a time, and take a 5 minute break. This technique allows astronauts to complete their work efficiently while also avoiding large amounts of work. This also allows astronauts to reduce their stress loads as it seems as they have less work to complete (Johnston). Another way astronauts learn efficiently is by using the feynman technique. This technique, created by Richard Feynman, discovered a way to make learning easy. Feynman believed teaching is the best way to learn. If you teach the subject you were trying to learn, you would learn it differently compared to just learning it normally. He says teach your subject as if you are teaching it to a child. This allows for you to learn your field in a unique way, and become a master of your technique.

Astronauts are tasked to be adaptable to any environment or situation they are put in. They are also needed to keep connections with family at home, keep in shape, using limited hygiene, establishing relationships with fellow crewmates and managing a good sleep schedule (CSA). These self management skills are vital as an astronaut. Keeping up with all given tasks can be stressful, that is where stress management comes into play (NASA).

Stress Management

Journaling as an astronaut is a must. Although you have crewmates to interact with, you miss your connections and relationships back at home. Journaling helps astronauts keep sane and avoids the urge of wanting to go back home (Soro). There are many psychological tricks and techniques astronauts use to keep themselves from being homesick and sad. Astronauts take breaks often to listen to music, read books, draw and do other things just like they would on earth. Mental health is important, it is crucial to take care of your emotional state before completing a task. If an astronaut is feeling down and is given a task, it is more than likely he or she will not complete it to the best of their ability. Crewmates are told to ask about others and care for one another, to ensure every crewmate's mental health is in check. Crewmates will often ask questions such as "How are you feeling today?", "What have you been up to?" and other general questions to get the crewmates to open up. Astronauts are also able to video call their family and talk to their friends through the phone to keep up with their social life on Earth.

Stress is a subject which tends to interfere with productiveness. To avoid having stress overcome your mental health, it is very important to try and take breaks as often as you can. Breaks reduce your stress load by a lot, and maintains a balance between your work and rest time.

Psychological Compatibility

Being psychologically strong is crucial. You must be able to adapt to fast paced environments without the need of being overwhelmed and taken over by stress. The compatibility and psychological requirements include being determined, taking initiative and being a leader as well as being a motivated person. The psychological aspect to being an astronaut is the most important detail companies look at before hiring.

Other skills required for an astronaut screening include good judgment, Integrity, reasoning, teamwork, communication, public speaking, motivation and resourcefulness. According to the CSA, Canadian citizens will be preferred rather than citizens of other countries.

You can argue that psychological compatibility is not as important compared to stress management. However, with a good mental state of mind to begin with, you are less prone to stress in the first place. Less stress leads to a higher productive state of mind, which gets work and tasks completed more efficiently at a higher rate. Psychological compatibility, or adaptation of the mental state of mind is a key factor of determining a productive individual from a nonproductive individual.

Conclusion

In conclusion, the standards of becoming an astronaut and going through astronaut screening are quite high. Years of dedication and hard work are needed prior to the astronaut screening to ensure a spot as an astronaut. You must become fit and strong, smart in all science fields, graduate with a bachelor's degree, have 20/20 vision, score a high IQ and much more. All that hard work just to have 2 people chosen in roughly 100 people chosen as candidates for astronaut screening. Even after successful reports and tests in astronaut screening, it all comes down to preference.

If you are thinking of becoming an astronaut in the future, this short chapter should be able to provide you with some context regarding a future job you may be considering. You may change your mind on becoming an astronaut after reading this, or strive harder to become one.

On the bright side, the future looks good for NASA. NASA hopes to send more people into space as technology continues to grow at an exponential rate. More astronauts will be chosen to be sent into space and you can expect space exploration to become much more advanced. You can also expect a higher chance of you becoming an astronaut in the future.

Chapter 7

Psychological Impacts and Preparation for Manned Missions

Brianna Bedran

Space Travel Stresses:

Space travel is an exciting adventure from what most of us have seen in media scaling from childhood tv shows to cinematic movies capturing the exhilarating moments of space travel. However, what occurs behind the scenes of being in a completely different environment away from planet earth can at times be far from joyful. Manned missions of space travel consist of severe isolation and confinement, communication difficulties, limited resources to address family emergencies, unfavourable team dynamics, heavy workloads, and on top of all this astronauts are not in a safe nor comforting environment. These severe stresses demonstrate behavioural and psychological changes for astronauts such as insomnia, fatigue, mood swings, withdrawal, and reduced motivation (Deming et al., 2017). Astronauts' average ratings of stress have been shown to be elevated pre- mission.Thus, countermeasures that boost psychological resiliency prior to mission launch are needed not only to mitigate stress during spaceflight but to address stressors confronted while preparing for spaceflight (Deming et al., 2017). Difficulties of addressing these problems arise as space travel is not commonly viewed from a holistic and occupation-based perspective (Davis, 2015) Isolation, confinement and induced stress factors also have a detrimental effect on cognitive and mental well-being, which could jeopardize mission accomplishment. Although countermeasures have been proposed, they mostly focus on cardiovascular and/or musculoskeletal systems. Long-term space flights require optimal cognitive performance of crew members during weightlessness for longer time periods independent of ground support (Frantzidis et al.,

2019). Living in a team with a risk of psychological incompatibility and the impossibility of urgent return to Earth add to the issue. A highly-trained medical person within the crew, diagnostic tools and equipment, psychophysiological support, countermeasures, and provisions for urgent, including surgical, treatment on board are necessary (Frantzidis et al., 2019).

Support:

Fortunately, although the field can improve psychological preparation for astronauts for manned missions, there are ways one can prepare a preflight mentality. Methods of support involve social support and workplace social support. This type of support recommended Charlene et al. addresses behavioral health outcomes such as better mood states, mental health disorders or symptoms, stress reactions, psychophysiological indicators of stress response (e.g., sleep quality), quality of life, well-being, resilience, posttraumatic growth, work attitudes, and work performance (Charlene et al., 2017).

Social support from within the workplace may regulate behavioral health concerns surfacing during the preflight period and can promote resiliency prior and during extended missions Support from others (i.e., social support) within the workplace can form the basis of preflight countermeasures for those entering, or potentially entering long-duration spaceflight. Social support countermeasures delivered within the workplace may be especially important in the context of limited access and isolation to other sources of social support (e.g., family and friends) during some stages of the preflight period (Charlene et al., 2017).

Social support may be categorized as: 1) emotional (i.e., care, trust, empathy, encouragement – and/or other positive emotional expressions); 2) informational (i.e., delivery of information facilitating problem-solving); 3) instrumental (i.e., concrete/tangible assistance); and 4) appraisal (i.e., delivery of information facilitating self-evaluation) support (Charlene et al., 2017).

Workplace social support as support provided from within a professional organizational context (e.g., military unit cohesion, mentoring, executive coaching), and, in relation to astronauts specifically, to NASA and/or from within the astronaut's broader professional community (e.g., retired astronauts) (Deming et al., 2017).

As described below under "Workplace Social Support Interventions," additional behavioral benefits may include: enhanced well-being, improved coping, greater perception of support, increased commitment to the organization, and enhanced job attitudes and performance, among other healthy outcomes (Deming et al., 2017). Trainees may realize that simply talking about hard experiences with mentors or colleagues can manifest supportive and structured leadership characteristics (e.g., informational support in the form of role clarity)8 as well as high levels of unit cohesion. Workplace Social Support Interventions also establish a safe environment of sharing for future astronauts (Deming et al., 2017).

It's just as important to have good quality of support in the field. In one study conducted for the mental preparation of flight for example, resulted in 89% of respondents indicated that limited time would prevent them from soliciting support from others, with other concerns including limits to confidentiality (68%), the potential for negative career impact (68%), and trouble accessing services (52%) (Deming et al., 2017). These findings suggest that stress may be associated with worsening perceptions of support from an organizational entity, constituting a potential barrier to delivery of social support unless anticipated and mitigated by the organization Observational findings by Charlene et al highlight significant relationships between workplace social support and behavioral health outcomes – both a positive impact in buffering against stress when workplace social support is present and a negative impact when there is a lack of workplace social support. These benefits include an enhanced well-being, increased ideas of interpersonal connection, better commitment to the organization, and improvement in job attitudes/performance (Deming et al., 2017). Further, research conducted on medical professionals and military personnel, who collectively share with astronauts attributes such as achievement orientation and mission focus, suggests that astronauts may both perceive the potential to benefit from workplace support and find social support delivered from within the workplace acceptable, especially when delivered by mentors and colleagues, as opposed to more structured organizational entities (Deming et al., 2017).

Psychological Preparation:

Understanding how space flight affects the human brain is essential to future space exploration. Several studies have reported anatomical and biochemical alterations, deterioration of the astronauts' performance and mental health, as well as disturbances in cortical activity and sleep. Significant factors contributing to impairments in alertness and performance during space missions, (Maiese, 2019) include sleep disturbances and alterations of circadian rhythms, as one night of sleep deprivation could deteriorate motivation, concentration, and increase cortisol levels. (Wu et al., 2018).

Ground-based simulators of microgravity are excellent mechanisms for preparing space flight experiments. They also facilitate stand-alone studies and cost-efficient platforms for gravitational research (Frantzidis, 2019) In space and bed rest simulations, there are a number of factors that may influence circadian rhythms and sleep. These include microgravity (gravitational force of, light flashes (light-dark cycles of approximately 90 minutes), light exposure (two-thirds of the time), low light intensity (below the threshold of efficiently entraining the human circadian clock), motion sickness, emotional stress, a high workload, an abnormal work/rest schedule, noise, thermal discomfort, muscle pain and confinement (Christos, 2019).

Limitations to ISSI simulations include: Not all experiments can be done in flight, there are limited resources (crew time, funds) available in flight and significantly longer times are required to complete studies (multiple flights are needed to achieve the required sample size). Thus why researchers recommend that ground-based analogues allow for the selection of the best candidate and countermeasures before testing them in flight. These analogues save time and money as studies can be completed more quickly and less expensively on the ground (Cromwell, 2018) Space flight analogues as used for human research create a situation that produces physiological and behavioural effects on the human body similar to those experienced in spaceflight. The analogue selection should take into consideration either the current ISS operations (Low Earth Orbit) or the future explorations (Moon, Mars) (Clement et al., 2019).

Research in the microgravity setting is indispensable to disclose the impact of gravity on biological processes and organisms. However, research in the near-Earth orbit is severely constrained by the limited number of flight opportunities. The term 'microgravity' is frequently used as a synonym for 'weightlessness' and 'zero-G,' which indicates that the G-forces are not actually zero but just 'very small'. Only true weightlessness can only be achieved in space (Christos, 2019).

Moreover, facilities are located in remote locations with extreme environments. Travel to these destinations can be difficult with little or no opportunity for emergency evacuation once the crew arrive. The goal of these missions is typically field research such as geological, environmental, or marine science. Human research is an adjunct to these missions. Therefore, crew responses to isolation and stress are studied (Christos, 2019).

Occupational Science:

Occupational science (OS) has been applied to the preparation and support of astronauts during long duration space exploration. Given the complex environment of space, it is not surprising that there is grave deterioration of both physical and mental health when off Earth. However, OS, through occupational therapy (OT), can identify strategies that maintain health and minimize disruptions in task performance for mission success (Davis, 2015).

Earth gravity has formulated cerebral functions related with motion, posture and navigation through a two-dimensional representation. However, weightlessness encountered within the International Space Station (ISS) and during space flights forces the brain to evolve in a three-dimensional (3D) environment. Reference points, due to specific gravitational constraints, thus do not exist anymore. The latter implies differences in posture and sensorimotor coordination. Weightlessness is combined with living within an extremely isolated environment. This is why it's recommended that crew members try to maintain an optimal status of cognitive functioning for performing complex mental tasks with circadian rhythm alterations (Davis, 2015).These factors may induce stress, sleep disorders and ;which contribute to detrimental effects on cognitive and mental well-being with the risk of jeopardizing inter-crew relationships (Davis, 2015) Space neuroscience primarily aims at 1) providing a better understanding of how reduced gravity affects brain functions, and 2) exploring countermeasures to ameliorate the detrimental effects of weightlessness to promote the success of space exploration missions (Davis, 2015).

Many OT practitioners recommended interventions that aligned with the gaps in knowledge identified by NASA on the Human Research Roadmap. For example, psychosocial interventions were mentioned only 52 out of 170 times (Davis, 2015). Occupational therapy practitioners employing psychosocial techniques can make further suggestions for closing this gap and identify and validate effective treatments for adverse social conditions and psychiatric annoyances that might occur during the mission. Specifically, interventions focusing on coping with stress, depression and other behavioral consequences, emotional regulation, mindfulness and meditation were frequently mentioned (Davis, 2015).

As mentioned, space travel is not commonly viewed from a holistic and occupation-based perspective. Occupational science studies are unique in that they sought to identify how Occupational Therapy (OT), with its foundation in Occupational Science (OS), could contribute to the well being and optimized task performance of astronauts. Occupational science is multidimensional with its philosophy predicated in a holistic view of the individual. The science applies studies of the components of function as well as occupations, encompassing the elements of cognition, psychomotor skills, emotions, social awareness, self-knowledge, and many other concepts associated with a variety of social science and health disciplines. Occupational science is primarily concerned with promoting health, well-being, and overall a higher quality of life through a balance of occupations. It is also used to inform all areas of OT practice (Henderson et al., 1991). By examining the transactional relationship between the individual (client factors) and skills needed for optimal task performance (performance skills), occupational therapy practitioners facilitate health, well-being, and enhanced participation in life's roles (The American Occupational Therapy Association, 2014).

OT interventions focused on stress management may provide opportunities for astronauts to deal with the dynamic interactions of mental, physical and psychological stressors associated with the astronaut experience. Occupational therapists prac- ticing in a psychosocial setting are highly skilled in helping clients overcome anxiety by teaching relaxation techniques and self regulation skills to recognize and manage their emotions, thoughts, and behaviors. Smith-Forbes et al. (Smith-Forbes, 2014) describes occupational therapy's role in addressing intellec- tual, physical, emotional and or behavioral reactions of service members in combat or other military operations. Stressors during combat may mirror.

It is exemplified in this chapter there are multiple forms of seeking help for psychological preparation and impacts of manned missions. One can turn to the workplace or social support, or seek help through more extensive means of psychoanalysis. However, the field must expand their means of providing any of this assistance for astronauts or anybody training for preflight. Most of the sources used identify that NASA has not been applying their best efforts to this type of mental preparation, and doing so can significantly improve astronauts motivation, reduce anxieties, and overall lead to more effective missions.

References

NASA. NASA health and medical policy for human space exploration. 2011a. [December 4, 2013]. (NPD 8900.5B). http://nodis3.gsfc.NASA.gov/npg_img/N_PD_8900_005B_/N_PD_8900_005B__main.pdf .

Johnston SL, Blue RS, Jennings RT, Tarver WJ, Gray GW. Astronaut medical selection during the Shuttle era: 1981–2011. Aviat Space Environ Med 2014; 85:823–7.

The NASA Twins Study: A multidimensional analysis of a year-long human spaceflight
Francine E. Garrett-Bakelman et al., Science, 2019

NASA astronaut Screening and testing,Felix A. Soto Toro, PH.D., PMP. 2012

Weight Loss in Humans in Space
Akiko Matsumoto et al., Aviation, Space, and Environmental Medicine, 2011

NASA. Astronaut candidate program. 2013a. [December 6, 2013]. http://astronauts.NASA.gov/content/broch00.htm .

NASA. NASA space flight human system standard. Volume 2: Human factors, habitability, and environmental health. 2011b. [December 4, 2013]. (NASA-STD-3001). https://standards.NASA .gov/documents/detail/3315785

IOM (Institute of Medicine). Review of NASA's space flight health standards-setting process: Letter report. Washington, DC: The National Academies Press; 2007.

Dezfuli H. Evolution of risk management at NASA and the philosophy of risk acceptance; PowerPoint presented at the second meeting of the Institute of Medicine Committee on Ethics Principles and Guidelines for Health Standards for Long Duration and Exploration Spaceflights; Washington, DC. July 25; 2013. [October 18, 2013]. http://www.iom.edu/~/media/Files/Activity%20 Files/Research/HealthStandardsSpaceflight/2013-JUL-25/Panel%20

1%20-%20Dezfuli.pdf . [PubMed]

IOM. Review of NASA's evidence reports on human health risks: 2013 letter report. Washington, DC: The National Academies Press; 2014. [PubMed]

NASA. NASA aviation medical certification standards. 2009a. [December 6, 2013]. (OCHMO 110902.2MED). http:// www.NASA.gov/pdf/620882main_Av_med_cert_std_ OCHMO_110902_2MED.pdf

NASA. Human integration design handbook (HIDH). 2010. [December 6, 2013]. (NASA/SP-2010-3407). http://ston.jsc.NASA .gov/collections/trs/_techrep/SP-2010-3407.pdf .

https://www.asc-csa.gc.ca/eng/astronauts/how-to-become-an-astronaut/requirements-and-conditions.asp

Washington, D.C. (2020). What Is Mental Health? MentalHealth. gov, Let's Talk About It. (Pg1)

Felman. A. (2020). What is mental health? MedicalNewsToday.

7activestudio. (2017, Mar 20). The Joy of Happy Learning.

The Royal Institution. (2015, Dec 19). A Place Called Space, Dr. Sarita Robinsons. [Video]. YouTube. https://m.youtube.com/ watch?v=o0rwGJ2b_U4&t=238s

Physiopedia. (2021). Salutogenic Approach to Wellness, 3. https:// www.physio-pedia.com/Salutogenic_Approach_to_Wellness.

NASA. Space Center Houston, Texas (2016, Apr 11). Evidence Report: Risk of Adverse Cognitive or Behavioral Conditions Psychiatric Disorders. Pg13.

Robinson J.N, Willis k, & Slack k. (2009 Jan). Patterns in crew-initiated photography of earth from ISS-Is earth observation a salutogenic experience? https://www.reserchgate.net.

Richard, M. & Edward, L. (2001). On Happiness and Human

Potentials: A Review of Research on Hedonic and Eudaimonic Well-Being. Pg143.

NOBASCHOLAR. (2018). Eudaimonia in the Contemporary Science of Subjective Well-Being: Psychological Well-Being, Self-Determination, and Meaning in Life. (Pg2). https://www.nobascholar.com. https://www.google.com/url?sa=t&source=web&rct=j&url.

Ryff, C. D. (2013). Eudaimonic Well-Being and Health: Mapping Consequences of Self Realization. In A. S. Waterman (E.d), the Best Within Us; Positive Psychology Perspectives on Eudaimonia (Pg72-98).

Huta, V. & Ryan, R. M (2010). Pursuing Pleasure or Virtue: The Differential and Overlapping Well-Being Benefits of Hedonic and Eudaimonic Motive. Journal of Happiness Studies. (pg11, 735).

The Free Dictionary. Psych- definition of psych. https://www.google.com/urlsa=t&source=web&rct=j&url.

USAFA Astronautics & Space Ops. (2020, Mar 15). Space Environment. [Video], YouTube. https://m.youtube.com/watch?v=snv1SnxcSbM&t=48s

The Space Environment. 4.1.2.2, The Space Environment and Spacecraft. https://www.google.com/url?sa=t&source=web&rct=j&url

NASA. (2015 July). Space Environmental Effects, Finckenor, M. M. & de Groh, K. K. https://www.google.com/url?sa=t&source=web&rct=j&url.

National Space Society. (2021, Jun 27). The Human Element: Surviving and Thriving in Space- Dr Chou, J. https://m.youtube.com/wattch?v=oZygeS8nANM&t=28s.

Australian Antarctic Program. Antarctic geology. Australian Government – Australian Antarctic Program. (2020, October 27). https://www.antarctica.gov.au/about-antarctica/geography-and-geology/geology/.

Bartone, P. T., Krueger, G. P., & Bartone, J. V. (2018). Individual differences in adaptability to isolated, confined, and extreme environments. Aerospace medicine and human performance, 89(6), 536-546.

Byers, M. (2020). Arctic Security and Outer Space. Scandinavian Journal of Military Studies, 3(1).

Cool Antarctic. (n.d.) Space Science in Antarctica. Astronomy and Astrophysics. Space Science in Antarctica, why is this a good place to study space from? https://www.coolantarctica.com/Antarctica%20 fact%20file/science/science-in-antarctica-space.php.

Dunbar, B. (2007). Greetings From the Bottom of the World. NASA. https://www.nasa.gov/audience/foreducators/k-4/features/F_ Greetings_From_the_Bottom_of_the_World.html.

Golden, S. J., Chang, C. H., & Kozlowski, S. W. (2018). Teams in isolated, confined, and extreme (ICE) environments: review and integration. Journal of Organizational Behavior, 39(6), 701-715.

LaFrance, A. (2015). The Worst Thing About Being in Space: Slow Internet. The Atlantic. https://www.theatlantic.com/technology/ archive/2015/06/the-internet-in-space-slow-dial-up-lasers-satellites/395618/.

National Aeronautics and Space Administration. Space environmental Effects. (n.d.) https://www.nasa.gov/sites/default/ files/files/NP-2015-03-015-JSC_Space_Environment-ISS-Mini-Book-2015-508.pdf

Pagel, J. I., & Choukèr, A. (2016). Effects of isolation and confinement on humans-implications for manned space explorations. Journal of Applied Physiology.

Palinkas, L. A. (2007). Psychosocial issues in long-term space flight: overview. Gravitational and Space Research, 14(2).

Perez, J. (2020). What Can We Learn About Isolation From NASA Astronauts? NASA. https://www.nasa.gov/feature/isolation-what-can-we-learn-from-the-experiences-of-nasa-astronauts.

Salazar, J. F. (2017). Antarctica and Outer Space: relational trajectories.

Temp, A. G., Lee, B., & Bak, T. (2020). "I really don't wanna think about what's going to happen to me!": a case study of psychological health and safety at an isolated high Arctic Research Station. Safety in Extreme Environments, 1-14.

Tortello, C., Barbarito, M., Cuiuli, J. M., Golombek, D. A., Vigo, D. E., & Plano, S. A. (2018). Psychological adaptation to extreme environments: Antarctica as a space analogue.

Zivi, P., De Gennaro, L., & Ferlazzo, F. (2020). Sleep in Isolated, Confined, and Extreme (ICE): A Review on the Different Factors Affecting Human Sleep in ICE. Frontiers in Neuroscience, 14.

Banks, S., Van Dongen, H. P., Maislin, G., &; Dinges, D. F. (2010). Neurobehavioral Dynamics Following Chronic Sleep Restriction: Dose-Response Effects of One Night for Recovery. Sleep, 33(8), 1013–1026. https://doi.org/10.1093/sleep/33.8.1013

Barger, L. K., Flynn-Evans, E. E., Kubey, A., Walsh, L., Ronda, J. M., Wang, W., Wright, K. P., &; Czeisler, C. A. (2014). Prevalence of sleep deficiency and use of hypnotic drugs in astronauts before, during, and after spaceflight: an observational study. The Lancet Neurology, 13(9), 904–912. https://doi.org/10.1016/s1474-4422(14)70122-x

Blodgett, R. (2018, January 16). Frequently Asked Questions. NASA. https://www.nasa.gov/feature/frequently-asked-questions-0/.

Buguet, A. (2007). Sleep under extreme environments: Effects of heat and cold exposure, altitude, hyperbaric pressure and microgravity in space. Journal of the Neurological Sciences, 262(1-2), 145–152. https://doi.org/10.1016/j.jns.2007.06.040

Czeisler, C., Bloomberg, J., &; Lee, A. (2002, December 16). Astronauts Need Their Rest Too: Sleep-Wake Actigraphy and Light Exposure During Space Flight - NASA Technical Reports Server (NTRS). NASA. https://ntrs.nasa.gov/citations/20030011396.

Frost Jr, J. D., Shumate, W. H., Salamy, J. G., & Booher, C. R. (1976). Sleep monitoring: the second manned Skylab mission. Aviation, space, and environmental medicine, 47(4), 372-382.

Germain, A., &; Kupfer, D. J. (2008). Circadian rhythm disturbances in depression. Human Psychopharmacology: Clinical and Experimental, 23(7), 571–585. https://doi.org/10.1002/hup.964

Government of Canada. (2019, May 16). Physical activity in space. Canadian Space Agency. https://www.asc-csa.gc.ca/eng/astronauts/living-in-space/physical-activity-in-space.asp.

Gundel, A., Nalishiti, V., Reucher, E., Vejvoda, M., &; Zulley, J. (1993). Sleep and circadian rhythm during a short space mission. The Clinical Investigator, 71(9). https://doi.org/10.1007/bf00209726

Guo, J.-H., Qu, W.-M., Chen, S.-G., Chen, X.-P., Lv, K., Huang, Z.-L., &; Wu, Y.-L. (2014). Keeping the right time in space: importance of circadian clock and sleep for physiology and performance of astronauts. Military Medical Research, 1(1). https://doi.org/10.1186/2054-9369-1-23

Kanas, N., & Manzey, D. (2008). Space psychology and psychiatry.

Koros, A., Wheelwright, C., & Adam, S. (1993). An evaluation of noise and its effects on shuttle crewmembers during STS-50/USML-1.

Kupfer, D. J. (1976). REM latency: a psychobiologic marker for primary depressive disease. Biological psychiatry, 11(2), 159-174.

Mallis, M. M., & DeRoshia, C. W. (2005). Circadian rhythms, sleep, and performance in space. Aviation, space, and environmental medicine, 76(6), B94-B107.

May, S. (2015, June 9). Sleeping in Space. NASA. https://www.nasa.gov/audience/foreducators/stem-on-station/ditl_sleeping.

Nasrini, J., Hermosillo, E., Dinges, D. F., Moore, T. M., Gur, R. C., &; Basner, M. (2020). Cognitive Performance During Confinement and Sleep Restriction in NASA's Human Exploration Research

Analog (HERA). Frontiers in Physiology, 11. https://doi.org/10.3389/fphys.2020.00394

Orzeł-Gryglewska, J. (2010). Consequences of sleep deprivation. International Journal of Occupational Medicine and Environmental Health, 23(1). https://doi.org/10.2478/v10001-010-0004-9

Palinkas, L. A. (2007). Psychosocial issues in long-term space flight: overview. Gravitational and Space Research, 14(2).

Sack, R. L., Auckley, D., Auger, R. R., Carskadon, M. A., Wright, K. P., Vitiello, M. V., &; Zhdanova, I. V. (2007). Circadian Rhythm Sleep Disorders: Part II, Advanced Sleep Phase Disorder, Delayed Sleep Phase Disorder, Free-Running Disorder, and Irregular Sleep-Wake Rhythm. Sleep, 30(11), 1484–1501. https://doi.org/10.1093/sleep/30.11.1484

Santy, P. A., Kapanka, H., Davis, J. R., & Stewart, D. F. (1988). Analysis of sleep on Shuttle missions. Aviation, space, and environmental medicine, 59(11 Pt 1), 1094-1097

Wu, B., Wang, Y., Wu, X., Liu, D., Xu, D., &p; Wang, F. (2018). On-orbit sleep problems of astronauts and countermeasures. Military Medical Research, 5(1). https://doi.org/10.1186/s40779-018-0165-6

Wulff, K., Gatti, S., Wettstein, J. G., &; Foster, R. G. (2010). Sleep and circadian rhythm disruption in psychiatric and neurodegenerative disease. Nature Reviews Neuroscience, 11(8), 1–23. https://doi.org/10.1038/nrn2868

Kanas, N., & Manzey, D. (2008). Space psychology and psychiatry (2nd ed). Microcosm Press ; Springer.

Kanas, N., Sandal, G., Boyd, J. E., Gushin, V. I., Manzey, D., North, R., Leon, G. R., Suedfeld, P., Bishop, S., Fiedler, E. R., Inoue, N., Johannes, B., Kealey, D. J., Kraft, N., Matsuzaki, I., Musson, D., Palinkas, L. A., Salnitskiy, V. P., Sipes, W., … Wang, J. (2009). Psychology and culture during long-duration space missions. Acta Astronautica, 64(7), 659–677. https://doi.org/10.1016/j.actaastro.2008.12.005

Oluwafemi, F. A., Abdelbaki, R., Lai, J. C.-Y., Mora-Almanza, J. G., & Afolayan, E. M. (2021). A review of astronaut mental health in manned missions: Potential interventions for cognitive and mental health challenges. Life Sciences in Space Research, 28, 26–31. https://doi.org/10.1016/j.lssr.2020.12.002

PhD, E. K. E. G. (1968). Mental Health Problems in Antarctica. Archives of Environmental Health: An International Journal, 17(4), 558–564. https://doi.org/10.1080/00039896.1968.10665281

Salamon, N., Grimm, J. M., Horack, J. M., & Newton, E. K. (2018). Application of virtual reality for crew mental health in extended-duration space missions. Acta Astronautica, 146, 117–122. https://doi.org/10.1016/j.actaastro.2018.02.034

Sandoval, L., Keeton, K., Shea, Christian, O., Patterson, L., & Leveton. (2021). Perspectives on Asthenia in Astronauts and Cosmonauts: Review of the International Research Literature.

Davis, J., Burr, M., Absi, M., Telles, R., & Koh, H. (2017). The contributions of occupational science to the readiness of long duration deep space exploration. Work, 56(1), 31–43. https://doi.org/10.3233/wor-162465

Deming, C. A., & Vasterling, J. J. (2017). Workplace Social Support and Behavioral Health Prior to Long-Duration Spaceflight. Aerospace Medicine and Human Performance, 88(6), 565–573. https://doi.org/10.3357/amhp.4778.2017

Christos F, Evangelia K, Aliki K, Vasilis N, Maria K and Christiane N. Current trends and future perspectives of space neuroscience towards preparation for interplanetary missions. (2019). Medknow Publications and Media Pvt. Ltd, 67 (8).

Clement G, Reschke MF. Neuroscience in Space. New York: Springer Science and Business Media; 2008. doi: 10.1007/978-0-387-78950-7. [Last accessed on 2019 Apr 23].

Cromwell R. (2018). Red Risk School: Space Flight Analogs [Presentation] NASA, Available from: https://media.bcm.edu/documents/2018/90/cromwell-red-risk-school-d5-space-flight-analogs-2.pdf. [Last accessed on 2019 Apr 23].

Henderson A, Cermak, S, Coster, W, Murray, E, Trombly, C, Tickle-Degnen L. (1991) The issue is: Occupational science is multidimensional. American Journal of Occupational Therapy ;45:370-2.

Maiese K. (2019). Impacting dementia and cognitive loss with innovative strategies: Mechanistic target of rapamycin, clock genes, circular non-coding ribonucleic acids, and Rho/Rock. Neural Regener Res;14:773-4.

Smith-Forbes ME, Najera C, Hawkins D. Combat oper- ational stress control in Iraq and Afghanistan: Army occupational therapy. Mil Med 2014;179(3):279-84.

Wu B, Wang Y, Wu X, Liu D, Xu D, Wang F. (2018) On-orbit sleep problems of astronauts and countermeasures. Mil Med Res; 5:1-12.

The American Occupational Therapy Association. (2014) The American Journal of Occupational Therapy:68(1): 1-52